UNIVERSITY OF CAPE COAST

FACULTY OF SCIENCE

DEPARTMENT OF CHEMISTRY

I0484920

PERFORMANCE EVALUATION OF KWANYAKO WATER TREATMENT PLANT

KWANYAKO HEADWORKS

BY

EMMANUEL KWAME ACKAH

DESERTATION PRESENTED TO FACULTY OF SCIENCE,

DEPARTMENT OF CHEMISTRY, UNIVERSITY OF CAPE COAST

IN PARTIAL FULFILMENT OF REQUIREMENTS FOR THE AWARD

OF BACHELOR OF SCIENCE (HONS) DEGREE IN

WATER AND SANITATION

MAY, 2012

DECLARATION

I, Ackah Emmanuel Kwame, do hereby declare that this project entitled "Performance

Evaluation of Kwanyako Water Treatment Plant, Kwanyako Headworks" was

exclusively done by me under the supervision of Mr. Philip Dwamena- Boateng. I further

declare, that this work herein presented is the result of my own research and investigation,

and that except for other people's work, which has been duly acknowledged, this

dissertation has not been presented to this University or elsewhere for any degree award.

....

Ackah Emmanuel Kwame   Mr. Philip Dwamena- Boateng

(Student)   (Supervisor)

Date..   Date..

i

DEDICATION

I dedicate this piece of research work to my parents, siblings and my wife.

ACKNOWLEDGEMENT

My foremost gratitude goes to NYAME SAFO for academic wit, physical and spiritual protection as well as favour and guidance throughout my education.

I also wish to express my profound gratitude to my parents especially, my wife for showing love, care, and being the brainwave and cornerstone of my tertiary education.

I sincerely thank and acknowledge the insight, edict, criticism and any form of assistance offered by my supervisor, Mr. Philip Dwamena-Boateng (Regional Production Manager, GWCL, Volta Region) during the execution of my research work.

Special thanks to the manager and staff of Quality Assurance Unit of GWCL, Kwanyako Headworks for their invaluable assistance without pliant.

My immense appreciation to my lecturers; Mr. Peter Appiah Obeng, Mr.Kwaku Oduru-Appiah, Mr. Albert Ebo Duncan Mrs. Martha Osei-Marfo all of UCC- Chemistry Department, Water and Sanitation Unit as well as Mr. Thomas Atteh Donkor of Ministry of Local Government and Rural Development for their advice and indiscrepancy behaviour during my period study.

I am extremely grateful to all my friends and loved ones especially, Stephen Aboagye and Daniel Maclar for happy and sad moments we have pooled together as great intimate associates without incongruity.

Last but not least, my mammoth gratitude to my entire programme mates for their harmony and resilience during the period of study.

ABSTRACT

The research was undertaken to evaluate the performance of Kwanyako Water Treatment Plant of GWCL. This is because the existing plant performance cannot meet the acceptable guidelines values of GDWS and WHO. The main objectives of the research were to determine the quality of raw and final water parameters at the headworks and compare the results to GDWS and WHO guidelines values. The parameters analysed include pH, colour, turbidity, residual chlorine, total hardness, chloride and iron. The results obtained from the analysis indicated that most parameters fell within GDWS and WHO guideline values. This showed that the new treatment plant now treats raw water to meet acceptable standards with few exceptions where in December, 2009 and 2010, the residual chlorine and colour of final water were below the acceptable guideline values. This may be due to seasonal variations and changes in raw water characteristics as a result of activities along the River Ayensu. Recommendations are that most economic activities along the river source should be halted and the chlorinator should be adjusted to yield considerable level of residual chlorine in finished water.

TABLE OF CONTENTS

Contents

LIST OF TABLES

LIST OF FIGURES

LIST OF ABBREVIATIONS

MDGs.. Millennium Development Goals

WHO... World Health Organisation

UNICEF... United Nations Internal Children's Education Fund

GWCL.. Ghana Water Company Limited

CWSA.. Community Water and Sanitation Agency

GDWS...Ghana Drinking Water Standards

PURC.. Public Utility and Regulatory Commission

DO... Dissolved Oxygen

TDS.................... ....Total Dissolved Solids

BOD.. Biochemical Oxygen Demand

COD.. Chemical Oxygen Demand

KWTP.. Kwanyako Water Treatment Plant

HU...Hazen Units

CHAPTER ONE

1. INTRODUCTION

1.1 Background to the Study

One of the most indispensable natural resources which cover about 70% of the earth surface is water. This resource is partly responsible and essential for the growth of all living organisms. Survival of these organisms also depends to large extent on water (www.importanceofwater.org/). About 10% of the total burden of diseases worldwide is due to unsafe water, sanitation and hygiene and consequently resulting in the loss of 3.6 million lives annually (Pruss-Ustun et al., 2008). Access to potable water and sanitation is important because it forms the basis of the health of communities and contributes significantly to health, economic and social gains (Montgomery et al., 2009). This presupposes that health is much dependent on the quality of water one drinks (Peavy & Rowe) and so, it is imperative therefore, that drinking water must be treated to meet certain standards before consumption. Irrespective of the gains made in the millennium development goals (MDGs) target of access to safe drinking water in the world, about 884million still do not have access to potable water supply and improved sanitation. Even with the current rate of progress, it has been estimated that 672million will still lack access to safe drinking water source by 2015; with more than one third coming from Sub-Saharan Africa. With reference to Ghana, 10% of the urban population still use unsafe sources of drinking water, whiles 26% of the rural population still use these (unsafe sources) of drinking water (WHO/UNICEF, 2010). The need to have safe drinking water in adequate quantities became known in the sixteenth century (Montgomery, 1985), beginning with public water supply in the 19th century

(Peavy S.H.et al,1985). Before then, the rationale of water purification was turbidity and better taste, not much was known about microorganisms or chemical contaminants (www.importanceofwater.org/). Public water supply began in Ghana in 1920s with a pilot pipe-borne water system in cape coast (GWCL, 2008). Presently, the urban water supply system is managed by Ghana Water Company Limited (GWCL) whiles rural water supply system is managed by the Community Water and Sanitation Agency (CWSA). The Ghana Water company operates 84 urban water supply systems throughout the country. The installed capacity of all the system is about 740,640m^3/day. Present potable water demand in the urban areas is estimated at about 1,101,032m^3/day, whiles average daily production is about 646,495m3/day. Effective urban coverage is therefore, about 59%; a coverage considered quite low compared with the United Nation's Millennium Development Goals and the Ghana's Poverty Reduction Strategy target (GWCL, 2008). Addressing these shortfalls, government of Ghana and other donor agencies reserves the sole responsibility of providing physical infrastructures that will support the provision of potable water on sustainable basis to meet the growing demand of the population. Integrated efforts are required through well designed and proper functioning of water treatment plants to meet the fast water supply demand. Therefore, water treatment plants (Headworks) are tasked with the objective of continuously producing adequate supply of water that protects and promotes the well-being of individuals and communities (M. E. Flentje, Water Quality and Treatment – Plant Control). This water from headworks must be wholesome (uncontaminated and free from excessive amount of mineral and organic matter), palatable and aesthetically acceptable (practically free of colour, turbidity, taste, odour and moderate temperature) drinking water. To achieve this, standards are set by GDWS to improving water quality and upholding stringent performance measures at the headworks. Similarly, the W.H.O has provided standards and guideline values of parameters for achieving quality water product to improving public health and socio-economic development of nations. These call for the establishment of quality control and assurance measures as well as facility management plan at the headworks to facilitate tracking down of plant records in order to improving plant performance and efficiency.

1.2   Problem Statement

The Old Kwanyako Water Treatment Plant was built in 1962 to supply water to the people of Kwanyako and its environs. Over the years, the quality of treated water produced from the plant could not meet acceptable standards of GDWS and W.H.O as a result of deterioration in raw water quality and inefficiencies in treatment. In attempt to resolving this problem, a new treatment plant was designed and built in 2007 and has been operating over the years. It is against this background that the research seeks to assess the performance of Kwanyako Water Treatment Plant.

1.3   Objectives of the Study

- ❖ To determine the quality of raw water at the headworks.
- ❖ To determine the quality of final water at the headworks.
- ❖ To compare the results with GDWS and W.H.O guideline values.
- ❖ To make recommendations.

1.4   Impact of Study

This research work will serve as a disclaimer to consumers' notion that the water produced by GWCL is of questionable quality if results prove otherwise. This will invariably raise the level of confidence of consumers in relation to patronage of water. Similarly, this research work will re-orient the public desire towards utility usage and imbue in them a justifiable pride of the treatment facility. It will reduce the use of sachet water tremendously, thereby eliminating the negative impact of plastic materials from sachet water on the environment. The incidence of many households resorting to the use of wells for various purposes if not eliminated will reduce drastically. However, the outcome of this research work would pave the grounds for civil society and other regulatory bodies such as PURC, Consumer Protection, etc. to exert pressure on utility providers to improve services if the outcome re-affirms claims by consumers.

1.5   Organization of Report

This report consists of five chapters. Chapter one deals with introduction which consists of background to the study, profile of the study area, objectives of the research and impact of the study as well as organization of report. Chapter two reviews all the literature and other document relevant to the study. Chapter three describes the methodology of the research work such as sampling procedure, equipment used and method of analysis. Chapter four gives a detail analysis of research results and discussions. Finally, chapter five presents the conclusions and recommendations of the research work.

CHAPTER TWO

2   LITERATURE REVIEW

2.1   Surface water quality

The quality of surface water depends on the catchment area drained, land use, location and source of natural and man-made pollution, and the natural agencies of purification such as sedimentation, sunlight, aeration, nitrification, filtration and dilution (Salvato, 1992). Water pollution may be defined as the presence of impurities in water in such a quantity and of such nature as to impair the use of the water for a stated purpose (Peavy *et al.*, 1985). Industrial activities, changing in agricultural practices and increasing urbanization can have significant influence on water quality (Tebbutt, 1998). Pollutants (impurities) accumulated by surface runoff through the hydrological cycle and as result of human activities may be in both suspended and dissolved form. Many parameters have evolved that qualitatively reflect the effect various impurities have on selected water uses. These parameters assess the physical, chemical and bacteriological characteristics of water (Anipa, 2001). Analytical procedures have been developed that quantitatively measure these parameters, and have been documented in 'Standard Methods for the Examination of Water and Wastewater' (Peavy *et al.*, 1985). These parameters have been classified into physical, chemical and bacteriological. The physical parameters, which include suspended solids, turbidity, colour, taste and odour, temperature, defines those characteristics of water that respond to the senses of sight, touch, taste or smell (Peavy *et al.*, 1985). Chemical parameters in water

5

quality include pH, alkalinity, hardness, dissolved oxygen, ammonia, organic nitrogen, nitrite, and chlorine and metals (toxic and non-toxic), etc. The bacteriological parameters define the occurrence of various bacteria, both pathogenic and non-pathogenic. The pathogenic organisms include vibrio cholera, salmonella typhi, streptococci, etc.

While these bacterial pathogens have been largely controlled by modern water treatment practices, other water-borne diseases can also be transmitted through the supply route, including those of viral and protozoan origin. Long established examples of diseases caused by these other groups include hepatitis, a disease due to a viral pathogen, and amoebic dysentery, a disease due to a class of protozoan pathogens (Bryant et al, 1992).

2.1.1 Standards and criteria

2.1.2 Drinking water standards

Standards of drinking water have evolved over the years with the growing knowledge of nature and effects of various contaminants. Minimum drinking water criteria established by W.H.O are used universally and are the basis for both EU and USA legislation. They have been established based on the following:

❖ Toxicology

❖ The feasibility of controlling the concentration of pollutants with known and economically available technology

❖ The ability to analyse for the contaminants (Hall et al, 1997).

❖ The Ghana Standard (GS 175) published in 1998 by the Ghana Standards Board specifies the physical, chemical, microbiological, biological and radiological requirements for drinking water.

It is generally essential that drinking water be of the following quality:

❖ free of suspended solids and turbidity

❖ be tasteless and odourless

❖ have only moderate quantities of dissolved inorganic solids

❖ not contain organics, toxics substances and pathogens (Peavy et al, 1985)

2.1.3 Water Treatment

A lot of sediments enter surface water together with other contaminants, making it turbid and unsafe for human consumption. Raw water therefore need to be treated so that it becomes free of pathogenic bacteria, aesthetically attractive and the final chemical content balanced to promote health make it potable for consumption purposes as well as ensuring adequate supply of water to meet the needs of the community at reasonable cost.

2.1.4 Water supply system

Every water supply system consists of a source of production (intake works), treatment plants and reservoirs, transmission lines and distribution networks. It is designed and constructed to supply a definite area called the Supply Area. It is defined during the design stage of the water supply scheme for a specific population projected over a designed period and the expected demand ((L.R., 1962).

2.1.5 Treatment Plant Control

The function of water treatment plant is to continuously produce an adequate water supply that protects and promotes the health and well-being of individuals and community. Such water is described as wholesome and palatable. Wholesome water is uncontaminated and free from excessive amounts of mineral and organic matter. A palatable and aesthetically acceptable drinking is free of colour, turbidity, taste and odour, and of moderate temperature (Boateng, 2012).Water treatment control requires constant operation and management of facilities personnel. Considerations must be given to factors such as competent personnel, standards of water quality plant maintenance and cleanliness, analytical laboratory control, measurement of water flow, chemical application control, operation and maintenance of chemical feeders and other plant equipment, storage and quality of chemicals, plant records and safety (Flentje, pp. 464-472).

2.1.6 Treatment Plant Appearance

The general appearance and cleanliness of the treatment plant as well as the grounds greatly influence the attitude of the public towards the utility. A clean and well-maintained plant will suggest a safe and potable product and competent management in which the public may have justifiable pride (Flentje M. , 1965). One of the greatest assets a public utility can earn is consumer confidence and goodwill. This can actually promote public health; for a community without confidence in its water supply may resort to the use of water from questionable sources. Additionally, it will also aid the plant operator to understand that he belongs to an industry in which cleanliness has important bearing on the quality of the finished product (Boateng, 2012).

2.1.7 Quality Specification for Finished Water

Each treatment plant should have a quality standard for its product. Plant control can become indefinite and uncertain procedure unless production is geared towards definite specifications of quality. The quality goals should be recognised by all personnel and constant effort made to adhere to these goals unless the water flowing through the plant is scheduled to receive treatment to achieve a predetermined end, the plant cannot be considered to be under proper control. The quality of finished water should be uniform (AWWA, 1968).

2.1.8 Water Treatment Processes

2.1.9 Storage

Raw water storage is an important aspect of the water treatment process. Storage is the process of purification which is carried out at impounded reservoir, basin for long period of time. Considerable time is required to accomplish this process, although bacteria and finely divided particles of are not entirely removed. The size of basins or reservoirs required makes this method costly. The tendency of stagnation and multiplication of low forms of animal and plant growth makes it undesirable. Though, many days of storage are provided, other methods of treatment are usually required to ensure satisfactory product (L.R., 1962). There is inevitable danger associated with prolonged storage, especially if the water remains stagnant which leads occasionally to high levels of growth. Algae growth in water storage causes fluctuations in oxygen concentrations.

2.1.10 Screening

This is necessary to remove floating debris such as dead wood logs, leaves, rags, etc. from water source, storm runoff events or snow melts. Some screens could be made of very fine materials or micro-strainers to remove suspended algae and plankton. As it is withdrawn from the source, surface water is usually screened through steel bars, typically about 1 inch (2.54 cm) thick and about 2 inch (5.08 cm) apart, to prevent large objects such as logs or fish from entering the treatment facility. Finer screens are sometimes employed to remove leaves. If the water is highly turbid (cloudy or muddy), it may be pre-treated in a large basin known as a pre-sedimentation basin to allow time for sand and larger silt particles to settle

2.1.11 Aeration

Aeration is the process of introducing air or oxygen into the water. In principle, four different types of aerators are common for gas transfer operations within the objectives of sanitary engineering. These are gravity (Cascades, Inclined planes, and vertical stacks), spray (Layout and Nozzle), air diffusers and mechanical aerators. The main purpose of aerator is to oxidize the iron and manganese. The various substances transferred into the raw water during aeration such as; oxygen, carbon dioxide, hydrogen sulphide, methane and volatile organic compounds are responsible for taste and odour. The process covers both physical removal of these undesirable gases as well as chemical removal by precipitation of dissolved minerals such as iron and to some extent manganese. There are a number of objectives to achieve during the aeration process;

❖ The addition of oxygen in water helps to oxidizes dissolved iron and manganese to insoluble forms which can then be removed by sedimentation and filtration. The equations below indicate the oxidation process:

1. $4Fe^{2+} + O_2 + 10H_2O \rightarrow 4Fe\,(OH)_3\downarrow + 8H^+$

2. $2Mn^{2+} + 2H_2O + O_2 \rightarrow 2MnO_2\downarrow + 4H^+$

❖ Removal of volatile organic compounds (suspected carcinogens)

❖ Removal of methane to prevent fire and explosions

❖ Removal of carbon dioxide to adjust or to approach the carbonate equilibrium with respect to calcium (Boateng, 2012)

2.1.12 Coagulation / Flocculation

Raw water abstracted from the source contains impurities such as suspended solids, colloidal particles and dissolved substances.

2.1.12.1 Suspended Solids

These are particle materials that consist of sand, silt, clays, etc. or organic products resulting from the decomposition of plant and animal matter, humic or formic acids. Microorganisms such as bacteria, plankton, algae and viruses are also found in suspended solids. These substances are particularly responsible for turbidity and colour of raw water.

2.1.12.2 Colloidal Particles

Colloidal particles also suspend in raw water because the particles range in size from less than one micron and have extremely slow settling rate. The particles subsequently cause turbidity and colour of raw water.

2.1.12.3 Dissolved Substances

These substances are soluble minerals usually anions or cations. Part of organic matter also dissolves in raw water. Dissolved gases such as oxygen, carbon dioxide and hydrogen sulphide subsequently produce associated odour.

Abstracted raw water is pumped into mixing basins where chemical coagulants such as aluminium sulphate (alum), sodium aluminate, ferric sulphate, ferrous sulphate, ferric chloride or polymers may be added. Coagulation is accomplished by adding chemical coagulants, usually alum and made effective by rapid mixing to disperse coagulants chemicals by violent agitation into the water being treated and flocculation to agglomerate small particles into well-defined flocs by gentle agitation for much longer time. When alum is placed in water, a chemical reaction occurs that produces positively charged aluminium ions. The overall result is the reduction of electrical charges and formation of sticky substance (flocs) which aggregate and eventually settles. Coagulation contributes to the removal of turbidity, colour and microorganisms (Duncan, 2011).

2.1.13 Sedimentation

Sedimentation is the process of separation of solid particles from suspension by the force of gravity. The process results in the removal of suspended solids and colloids through settling of particles with minimum turbulence at entry and exits points with minimum short-circuiting in the clarifiers. Typical basins used in sedimentation include conventional rectangular basins, conventional centre-feed basins, peripheral-feed basins and spiral-flow basins. Sludge accumulates at the bottom of the basin. In conventional water treatment, detention time required for settling depends largely on total filter capacity (Howard S. P., 1985).Various types of particles which settle out include:

2.1.13.1 Discrete settling

In this type of settling, particle size, shape and specific gravity do not change with time. Usually occurs in plain sedimentation.

2.1.13.2 Flocculent settling

Here, particles aggregate or coalesce with other particles upon contact, thus changes in shape, size and perhaps specific gravity occur.

2.1.13.3 Zone settling

It is the settling of suspension intermediate in concentration between discrete settling and compression (settling). The particles are so close together that inter-particle forces hinder settling of neighbouring particles. The particles remain in fix position relative to each other and all settle at constant velocity

2.1.13.4 Compression

This is the settling of particles at very high concentrations as would occur near the bottom of settling basin. The particles actually touch each other and settling can occur by compression of the compacting mass

2.1.14 Filtration

Filtration is a physical process of separating suspended and colloidal particles from water by passing the water through a granular material. The process of filtration involves straining (where particles are captured in the small spaces between filter media grains), settling (where particles land on top of the grains and stay there) and adsorption (where a chemical attraction occurs between the particles and the media grains (Spellman, 1999).Filters have been found to be effective for the removal of particulate of all size ranges such as algae, colloidal humic compounds, viruses, asbestos fibres and colloidal clay particles, bacteria, protozoa, etc., provided the appropriate design parameters are used. Though, it is still advisable to use disinfectants as precautionary measure (James, 1985).

2.1.14.1 Types of filters

Different types of filters are available for use in water treatment plants and are classified as follows:

❖ Hydraulically as rapid or slow, depending on the rate of flow per unit of surface area;

❖ According to the kind of or type of filter media employed as sand, coal (or anthracite), coal sand, single, dual or multi-layered mixed bed, diatomaceous bed, etc.

❖ According to the direction of flow through the bed as down flow, up flow, bi-flow, fine-to-coarse, or coarse-to-fine;

❖ As gravity or pressure depending on the mode of filtration;

❖ According to the method of flow control as constant rate or declining rate (ASCE & CSSE, (1971).)

❖ In filters, impurities from the raw water are removed by a combination of different processes such as sedimentation, straining, adsorption, biochemical and microbial activities (Buamah, (2010))

2.1.14.2 Adsorption

The use of solids for removing undesirable substances from either gaseous or liquid solution has been widely used for long time. This process, known as adsorption (Masel, (1996).), involves the preferential concentration of substances from the gaseous or liquid phase onto the surface of a solid substrate. Adsorption is operative in most natural physical, biological and chemical systems, employing

14

solids such as activated carbon and synthetic resins as adsorbent for industrial application and purification of water among others (http://www.seperationprocesses.com).

In water purification processes, adsorption in rapid filtration systems remove finely divided suspended matter as well as colloidal and molecular dissolved impurities. The forces of adsorption exert their influence over short distances only between 0.01-0.1 um (Huisman, (1986)) (Sharma, (1997)) and the water film surrounding the filter grains has much greater thickness. It has been well established that, there are two types of adsorption, namely Physical adsorption and Chemical adsorption (http://www.seperationprocesses.com).

2.1.14.3 Physical adsorption

Physical adsorption is caused mainly by weak van der Waals forces and to some extent by electrostatic forces between adsorbate and adsorbent surface. The adsorbed material is not rigidly fixed to a specific site at the surface hence is free to undergo transitional movement within the interface. Thus adsorbed material may condense and form several superimposed layer on the surface of the adsorbent. Consequently, physical adsorption is generally quite reversible (http://www.seperationprocesses.com).

2.1.14.4 Chemical adsorption

Chemisorption, in contrast, involves relatively strong forces between the adsorbed molecules and the surface. As a result, the adsorbate is not free to move on the surface. Chemisorption often forms monolayer (Sharma, (1997))

2.1.15 Mechanism of adsorption

A solid in contact with a liquid is covered by a stagnant film of liquid in which reactants have to diffuse before they reach the interface to react. The interaction between a solid and a liquid therefore, takes place through the following steps:

❖ Diffusion of reacting molecules to the interface

❖ Reaction at the interface

❖ Desorption of products

❖ Adsorption at the interface

❖ Diffusion of products from the interface (Mormade, 2001).

2.1.16 Disinfection

Disinfection is a process used to control waterborne pathogenic organisms and prevent waterborne diseases. The major goal of disinfection in water system is to destroy all disease-causing organisms (Spellman, 1999). The effectiveness of disinfection in a drinking water system is measured by testing for the presence or absence of coliform bacteria. Their presence indicates possibility of contamination and their absence suggests possibility that the water is potable if source is adequate, headworks history is good and acceptable residual chlorine is present (Peavy & Rowe). The desired characteristics of a disinfectant include the following:

❖ It must be able to deactivate any type or number of disease-causing microorganisms that may be in water supply, in reasonable time, within expected temperature range and changes in the character of the water.

❖ It should persist within the disinfected water at high enough concentration to provide residual protection through the distribution.

❖ It must be nontoxic

❖ It must not add unpleasant taste or odour to the water.

❖ It must be quick and easy to determine the concentration of the disinfectant in the treated water (Spellman, 1999). Disinfectants are also used to achieve other specific objectives in drinking water treatment. These objectives include nuisance control, oxidation of specific compounds and use as a coagulant and filtration aid. The most commonly used disinfectants and oxidants in no particular order are chlorine, chlorine dioxide, chloramines, ozone and potassium permanganate.

2.1.17 Softening (pH Adjustment)

Softening is a process used to reduce hardness in water treatment. Hardness is mainly caused by the presence of calcium and magnesium metallic charged irons in solution in the water. The length of time that water gets in contact with hardness-producing material is a major factor that determines how much hardness there is in the raw water. Generally, two types of methods are used for softening, namely 'lime-soda ash' process and 'cation-ion exchange' process. To soften water by the lime-soda ash method, its degree of alkalinity has to be considered. The alkalinity of a water sample is a measure of the water's capacity to neutralize acids. In natural and treated waters, alkalinity is the result of the presence of bicarbonates, carbonates and hydroxides of calcium, magnesium and sodium. Chemicals used in water treatment such as alum, chlorine or lime cause change in alkalinity. Alkalinity is usually express in terms of calcium carbonate.

Alkalinity = bicarbonate ion conc. (HCO_3^-) + carbonate ion conc. (CO_3^-) + hydroxyl ion conc. (OH^-), expressed as calcium carbonate ($CaCO_3^-$). Chemical precipitation is one of the most common methods used to soften water. Chemicals normally used are lime [$Ca(OH)_2$] and soda ash ($NaCO_3$). Lime is used to remove the chemical that causes temporal (carbonate) hardness while soda ash is used to remove permanent (non-carbonate) hardness. When lime and soda ash are added, the hardness-causing mineral formed is nearly insoluble precipitates. When calcium hardness is removed from a chemical softener, it is precipitate as calcium carbonate while magnesium hardness is removed from chemical softener; it is precipitated as magnesium hydroxide.

2.2 WATER QUALITY PARAMETERS

Water quality focuses on the presence of foreign substance in water and their effect on people and the environment. Water of good quality for one purpose may be considered to be poor quality for some other use. As such, protecting water quality and modifying it for a particular purpose are the major objectives in the field of water treatment (J.A., 2000)

2.2.1 pH

pH is an important limiting chemical factor for aquatic life. If the water in a stream is too acidic or basic, the H+ or OH- ion activity may disrupt aquatic organism's biochemical reactions by either harming or killing the stream organisms. pH is expressed in a scale with ranges from 1 to 14. A solution with a pH less than 7 has more H+ activity than OH-, and is considered acidic. A solution with a pH value greater than 7 has more OH- activity than H+, and is considered basic. The pH scale is logarithmic, meaning that as you go up and down the scale, the values change in factors of ten. A one-point pH change indicates the strength of the acid or base has increased or decreased tenfold. Streams generally have a pH values ranging between 6 and 9, depending upon the presence of dissolved substances that come from bedrock, soils and other materials in the watershed. Changes in pH can change the aspects of water chemistry. For example, as pH increases, smaller amounts of ammonia are needed to reach a level that is toxic to fish. As pH decreases, the concentration of metal may increase because higher acidity increases their ability to be dissolved from sediments into the water (Streamkeeper's Field Guide, 1991). Hardness is frequently used as an assessment of the quality of water supplies. The hardness of a water is governed by the content of calcium and magnesium salts (temporary hardness), largely combined with bicarbonate and carbonate and with sulfates, chlorides, and other anions of mineral acids (permanent hardness) (Whipple, 1947).

2.2.2 Alkalinity

The Alkalinity or the buffering capacity of a stream refers to how well it can neutralize acidic pollution and resist changes in pH. Alkalinity measures the amount of alkaline compounds in the water, such as carbonates, bicarbonates and hydroxides. These compounds are natural buffers that can remove excess hydrogen (H+) ions (Standard Methods 19th Edition, 1995).

2.2.3 Dissolved Oxygen

The amount of Dissolved Oxygen (DO) in water is expressed as a concentration. A concentration is the amount of in weight of a particular substance per a given volume of liquid. The DO concentration in a stream is the mass of the oxygen gas present, in milligrams per liter of water. Milligrams per liter (mg/L) can also be expressed as parts per million (ppm). Dissolved Oxygen (DO) is essential for aquatic life. A low DO (less than 2mg/l) would indicate poor water quality and thus would have difficulty in sustaining many sensitive aquatic lives. The concentration of dissolved oxygen in a stream is affected by many factors:

❖ Temperature: Oxygen is more easily dissolved in cold water.

❖ Flow: Oxygen concentrations vary with the volume and velocity of water flowing in a stream. Faster flowing white water areas tend to be more oxygen rich because more oxygen enters the water from the atmosphere in those areas than in slower, stagnant areas.

❖ Aquatic Plants: The presence of aquatic plants in a stream affects the dissolved oxygen concentration. Green plants release oxygen into the water during photosynthesis. Photosynthesis occurs during the day when the sun is out and ceases at night. Thus in streams with significant populations of algae and other aquatic plants, the dissolved oxygen concentration may fluctuated daily, reaching its highest levels in the late afternoon. Because plants, like animals, also take in oxygen, dissolved oxygen levels may drop significantly by early morning.

❖ Altitude: Oxygen in more easily dissolved into water at low altitudes that at high altitudes.

❖ Dissolved or suspended solids: Oxygen is also more easily dissolved into water with low levels of dissolved or suspended solids per litre of water. Milligrams per litre (mg/L) can also be expressed as parts per million (ppm). Dissolved Oxygen (DO) is essential for aquatic life. A low DO (less than 2mg/l) would indicate poor water quality and thus would have difficulty in sustaining many sensitive aquatic lives. The concentration of dissolved oxygen in a stream is affected by many factors:

❖ Temperature: Oxygen is more easily dissolved in cold water.

❖ Flow: Oxygen concentrations vary with the volume and velocity of water flowing in a stream. Faster flowing white water areas tend to be more oxygen rich because more oxygen enters the water from the atmosphere in those areas than in slower, stagnant areas.

❖ Aquatic Plants: The presence of aquatic plants in a stream affects the dissolved oxygen concentration. Green plants release oxygen into the water during photosynthesis. Photosynthesis occurs during the day when the sun is out and ceases at night. Thus in streams with significant populations of algae and other aquatic plants, the dissolved oxygen concentration may fluctuated daily, reaching its highest levels in the late afternoon. Because plants, like animals, also take in oxygen, dissolved oxygen levels may drop significantly by early morning.

❖ Altitude: Oxygen in more easily dissolved into water at low altitudes that at high altitudes. Dissolved or suspended solids: Oxygen is also more easily dissolved into water with low levels of dissolved or suspended solids. Usually streams with high dissolved oxygen concentrations (greater than 8 mg/L for Ozark streams) are considered healthy streams. They are able to support a greater diversity of aquatic organisms. They are typified by cold, clear water, with enough riffles to provide sufficient mixing of atmospheric oxygen into the water (Institute of Water Engineers, 1954).

2.2.4 Colour (Hazen)

Colour is vital as most water users, be it domestic or industrial, usually prefer colourless water. Determination of colour can help in estimated costs related to discolouration of the water.

21

2.2.5 Conductivity

Conductivity indicates the presence of ions within the water, usually due to in majority, saline water and in part, leaching. It can also indicate industrial discharges. The removal of vegetation and conversion into monoculture may cause run-off to flow out immediate thus decrease recharge during drier period. Hence, saline intrusion may go upstream and this can be indicated by higher conductivity (AWWA, 1968).

2.2.6 Turbidity (NTU)

Turbidity may be due to organic and/or inorganic constituents. Organic particulates may harbour microorganisms. Thus, turbid conditions may increase the possibility for waterborne disease. Nonetheless, inorganic constituents have no notable health effects. The series of turbidity-induced changes that can occur in a water body may change the composition of an aquatic community. First, turbidity due to a large volume of suspended sediment will reduce light penetration, thereby suppressing photosynthetic activity of phytoplankton, algae, and macrophytes, especially those farther from the surface. If turbidity is largely due to algae, light will not penetrate very far into the water, and primary production will be limited to the uppermost layers of water. Cyanobacteria (blue-green algae) are favoured in this situation because they possess flotation mechanisms. Overall, excess turbidity leads to fewer photosynthetic organisms available to serve as food sources for many invertebrates.

As a result, overall invertebrate numbers may also decline, which may then lead to a fish population decline. If turbidity is largely due to organic particles, dissolved oxygen depletion may occur in the water body. The excess nutrients available will

22

encourage microbial breakdown, a process that requires dissolved oxygen. In addition, excess nutrients may result in algal growth. Although photosynthetic by day, algae respire at night, using valuable dissolved oxygen. Fish kills often result from extensive oxygen depletion (US EPA, 1998).

2.2.7 Salinity

High salinity may interfere with the growth of aquatic vegetation. Salt may decrease the osmotic pressure, causing water to flow out of the plant to achieve equilibrium. Less water can be absorbed by the plant, causing stunted growth and reduced yields. High salt concentrations may cause leaf tip and marginal leaf burn, bleaching, or defoliation. As per Conductivity, salinity (NaCl content, g/kg) can be used to check for possible saline intrusion in future (Standard Methods 19th Edition, 1995)

2.2.8 Total Suspended Solid

A total Suspended solid is an indication of the amount of erosion that took place nearby or upstream. This parameter would be the most significant measurement as it would depict the effective and compliance of control measures e.g. riparian reserve along the waterways. The series of sediment-induced changes that can occur in a water body may change the composition of an aquatic community. First, a large volume of suspended sediment will reduce light penetration, thereby suppressing photosynthetic activity of phytoplankton, algae, and macrophytes. This leads to fewer photosynthetic organisms available to serve as food sources for many invertebrates. As a result, overall invertebrate numbers may also decline, which may then lead to decreased fish populations. In addition, sediment may interfere with essential functions of organisms. The numbers of filter-feeding invertebrates will decline if their filter mechanisms are choked by suspended particles. Some zooplankton suffers due to clogged feeding mechanisms. Likewise, fish may suffer clogging and abrasive damage to gills and other respiratory surfaces. Abrasion of gill tissues triggers excess mucous secretion, decreased resistance to disease, and a reduction or complete cessation of feeding. Suspended sediment may also affect predator-prey relationships by inhibiting predators' visual abilities (American Public Health Association, 1965).The settling of suspended solids from turbid waters threatens benthic aquatic communities. Deposited particles may obscure sources of food, habitat, hiding places, and nesting sites. Most aquatic insects will simply drift with the current out of the affected area. Benthic invertebrates that prefer a low-silt substrate, such as mayflies, stoneflies, and caddis flies, may be replaced by silt-loving communities of oligochaetae, pulmonate snails, and chironomid larvae. Increased sediment may impact plant communities. Primary production will decline because of a reduction in light penetration. Sediment may damage plants by abrasion, scouring, and burial. Finally, sediment deposition may encourage species shifts because of a change of substrate. Sediment deposition may also affect the physical characteristics of the stream bed. Sediment accumulation causes stream bed elevation and a decrease in channel capacity. Flooding is more likely after sediment accumulation because the stream cannot accommodate the same volume of water. Also, a substrate that is closer to the surface receives more light and supports increased numbers of photosynthetic organisms, such as rooted algae. As a result, recreational use may be threatened because moving parts of boats may become tangled in aquatic plants. Sediment, which is generally negatively charged, attracts positively charged molecules. Some of these molecules (phosphorus, heavy metals, and pesticides) are pollutants. These positively charged pollutants are in equilibrium with the water column and are often released slowly into the water resources (Merrill, 1962).

2.2.9 Total Dissolved Solids (TDS)

The total dissolved solids (TDS) in water consist of inorganic salts and dissolved materials. In natural waters, salts are chemical compounds comprised of anions such as carbonates, chlorides, sulphates, and nitrates (primarily in ground water), and cations such as potassium (K), magnesium (Mg), calcium (Ca), and sodium (Na). In ambient conditions, these compounds are present in proportions that create a balanced solution. If there are additional inputs of dissolved solids to the system, the balance is altered and detrimental effects may be seen. Inputs include both natural and anthropogenic source (Buswell, 1938).

2.2.10 Biochemical Oxygen Demand,

BOD is a measure of organic pollution to both waste and surface water. High BOD is an indication of poor water quality. For this tree plantation project, any discharge of waste into the waterways would affect the water quality and thus users downstream (Buswell, 1938).

2.2.11 Chemical Oxygen Demand, COD

COD is an indicator of organics in the water, usually used in conjunction with BOD. High organic inputs trigger deoxygenation. If excess organics are introduced to the system, there is potential for complete depletion of dissolved oxygen. Without oxygen, the entire aquatic community is threatened. The only organisms present will be air- breathing insects and anaerobic bacteria. If all oxygen is depleted, aerobic decomposition ceases and further organic breakdown is accomplished anaerobically. Anaerobic microbes obtain energy from oxygen bound to other molecules such as sulphate compounds. Thus, anoxic conditions result in the mobilization of many otherwise insoluble compounds. In areas of high organics there is frequently evidence of rapid sewage fungus colonization. Sewage fungus appears as slimy or fluffy cotton wool-like growths of micro-organisms which may include filamentous bacteria, fungi, and protozoa such as Sphaerotilusnatans, Leptomitus lacteous, and Carchesium polypinuym, respectively. The various effects of the sewage fungus masses include silt and detritus entrapment, the smothering of aquatic macrophytes, and a decrease in water flow velocities. An accumulation of sediment allows shifts in the aquatic system structure as colonization by silt-loving organisms occur. In addition, masses of sewage fungus may break off and float away, causing localized areas of dissolved oxygen demand elsewhere in the water body. Organic levels decrease with distance away from the source. In a standing water body such as a lake, currents are generally not powerful enough to transport large amounts of organics. In a moving water body, the saprotrophic organisms (organisms feeding on decaying organic matter) break down the organics during transportation away from the source. Hence, there is a decline in the oxygen demand and an increase of dissolved oxygen in the water. Community structure will gradually return to ambient with distance downstream from the source (Streamkeeper's Field Guide, 1991)

2.2.12 Nitrate Nitrogen

The growth of macrophytes and phytoplankton is stimulated principally by nutrients such as nitrates. Many bodies of freshwater are currently experiencing influxes of nitrogen and phosphorus from outside sources. The increasing concentration of available phosphorus allows plants to assimilate more nitrogen before the phosphorus is depleted. Thus, if sufficient phosphorus is available, high concentrations of nitrates will lead to phytoplankton (algae) and macrophyte (aquatic plant) production. This is mostly due to the usage of fertilizers (M.D, 1985).

2.2.13 Electrical conductivity

The electrical conductivity (EC) of water sample is its ability to conduct an electric current. EC is actually a measure of the ionic activity of a solution in terms of its capacity to transmit current. It is an extremely useful measurement used to estimate the concentrations of dissolved inorganic solids (ions) in water. Thus EC is related to T.D.S. The conductivity is related to the total concentration of ions in solution, their valence mobility and to the temperature of measurement. It is more related to inorganic components of the sample may signal changes in mineral composition of raw water, seasonal variations in reservoirs, intrusion of sea or saline water by over pumping and pollution from industrial waters. Since ions have different motilities, the relation between total dissolve solids and conductivity is not constant for all waters but varies in range as; conductivity (us/cm, or micro ohms/cm) x factor (0.55 to 0.99) = T.D.S mg/L [1 μs/cm =1 μohm/cm]. The variation of conductivity with temperature is an increase of 1.9

per °c. Therefore, it is important to measure the temperature of the sample while measuring its conductivity (Joanne E. Drinan, 2009).

2.2.14 Total Hardness

Hardness is defined as the amount of the concentrations of calcium and magnesium ions dissolved in water. These two ions are the major hardness constituents and though some other metals contribute to hardness, their concentrations in natural waters are so much smaller that their significance as hardness is negligible. Calcium is the most abundant dissolved atomic constituent of natural freshwaters and is widely distributed in the minerals of rocks and soils. It is the fifth most abundant element on the earth and is found in every major areas of the world. Magnesium is also a most constituent of rocks, in abundance second to calcium and is usually found in the occurrence with calcium. Hard waters leave spot on glasses dingy film on laundry and hair and crusty deposits on bathroom fixtures, the presence of hardness in water supplies also contribute to taste, odour, colour or turbidity to the water. The carbonate salts of calcium are the major source of dissolved calcium and are generally referred to as limestone or calcite. They include ice land spare (pure), marble and alabaster (less pure and more compressed), and chalk. Calcium carbonate is quite insoluble in water and dissolves only up to 15mg/L but if CO_2 is present, this natural acidity makes limestone much more soluble. This often occurs in groundwater since bacterial action in the soil releases CO_2 changing carbonate to bicarbonate and dissolving large amounts of calcium into water. Because of this groundwater are generally harder waters than surface waters (American Public Health Association, American Water Works Association and Water Environment Federation, 1998).

28

2.2.15 Calcium Hardness

Concentration of calcium is routinely measured separately from total hardness. It concentration in water can range from zero to thousand mg/L as $CaCO_3$ (American Water Works Association, 1995).

2.2.16 Magnesium Hardness

Magnesium is also routinely determined by subtracting calcium hardness from total hardness. There is usually less magnesium than calcium in natural water. Lime dosage for water softening operation is partly based on the concentration of magnesium hardness in water. (AWWA, 1995).

2.2.17 Chloride

The origin of chloride in groundwater is more complex since increased chloride values could be due to trapped fossil sea water, leaching of evaporates deposits or even washing out of chloride from overlying rocks and soils. If penetration by deep drilling, artesian saline water can leak into upper fresh water aquifers. The effects of chloride on taste may be critical to its use. Some water containing 250mg/L of chloride may have a detectable salty taste if the predominant cation is sodium but if the predominant cation is calcium or magnesium the salty taste may not be apparent even up to 1000mg/L of chloride. The guide level of chloride according to WHO (1983) is 250mg/L for drinking water on organoleptic considerations. Chlorides accelerate corrosion of metals. (Lonore S. Chassed, Arnold E. Greenberg, and Andrew D. Eaton, 1998).

2.2.18 Manganese

In natural water manganese is usually found co-existing with irons, but is more of a problem because its oxide deposits are black, denser and often more difficult to remove. In groundwater, it usually exists as the Mn^{2+} ion due to lack of oxygen but soon precipitate black, mixed oxides on exposure to air. Levels greater than the WHO guide level of 0.1mg/L are sufficient to cause staining, washing and taste problems. In some process such as paper making, dying and beverage manufacturing, the levels must not exceed 0.02mg/L (K.S. Venkateswarlu, 1996).

2.2.19 Microbiological

2.2.19.1 *Faecal Coliform Count*

Microbiological test is to detect the Level of pollutions caused by living thing especially human who live or work in the area especially upstream of the site. These tests are based on coliform bacteria as the indicator organism. The presence of these indicative organisms is evidence that the water has been polluted with faeces of humans or other warm-blooded animals (Standard Methods 19th Edition, 1995).

2.2.19.2 Pesticides

These parameters are common tests for the level of agrochemical pollution. Since a specific type of agrochemical to use is unknown at this stage, it is unknown at this stage the type of agrochemical that would be used in the proposed development, a range of test is recommended for analyzing to gauge the existing condition that could be used as baseline information or reference (US EPA,1996).

2.2.20 Record Keeping

The following are the basic records that must be maintained:

❖ The date of each cleaning (commencement)

❖ The date and hour of return to full service (end of re-ripening period)

❖ Raw and filtered water levels (measured each day at the same hour) and daily loss of head. The filtration rate, the hourly variations, if any.

❖ The quality of raw water in physical terms (turbidity, colour) and bacteriological terms (total bacterial count, E. Coli.) determined by samples taken each day at the same hour.

❖ The same quality factors of the filtered water.

❖ Any incidents occurring e.g. plankton development, rising Schmutzdecke, and unusual weather conditions.

CHAPTER THREE

3   METHODOLOGY

3.1   REARCH DESIGN

The project work was designed to assess (determine)whether the quality of water supplied in and around the catchment area of Kwanyako Water Treatment Plant (KWTP) meets the World Health Organization (W.H.O) and Ghana Drinking Water Standard (GDWS) guidelines.

3.2   Profile of Kwanyako Water Treatment Plant

Kwanyako Water Treatment Plant with present capacity of 35000 m^3/d (7,700,000 gal/d) is located 10km east of Agona Swedru and abstract raw water from the Ayensu River, a medium – sized river which flows almost centrally through the supply area of the Kwanyako water supply system. Kwanyako Headworks consist of two treatment plants; the old plant and the Jubilee plant (new plant). The former was commissioned on the 25th January, 1964 by Hon. E.K. Bansah, MP and Minister of works and Housing. The old plant with the capacity of 12,440 m^3/d (2,728,000 gal/d) serves eight (8) districts all in the Central Region of Ghana with population greater than 5,000 and some 300 surrounding villages with its pipeline estimated over 120km. The later with a capacity of 21,000m^3/d (4,620,000 gal/d) was commissioned on 21st February, 2007 by H.E. John Agyekum Kuffuor, former President of the Republic of Ghana. Kwanyako Water Treatment Plant was designed as conventional treatment plant which undergoes screening, aeration, coagulation, flocculation, filtration, disinfection, pH correction and wastewater treatment (Slow Sand Filter) although water is allowed to flow through the entire treatment unit originally provided, such as clariflocculators. (Anor G.A, 2012). Raw water is abstracted from

32

the dam at the intake point and flows by gravity to the six (6) number vertical submersible pumps from where it is moved through a height of 35m to the one (1) number aerator at the treatment works located about a kilometre (1 km) from the dam. The raw water is exposed to the atmosphere at the aerator by cascading for removal of dissolved iron, manganese and odour in the water while oxygen is introduced. Alum is added and the water flows by gravity through a dividing chamber to the four (4) clariflocculators where suspended matter is flocculated and resettled. The clarified (settled) water then flows to the twelve (12) numbers rapid gravity sand filters for the removal of residual floccules (flocs) for clear water emerge. Chlorination, a process where chlorine is dosed for the removal of algae and arrest of any biological growth is performed before lime is added for pH adjustment and correction. The water then flows into a disinfection contact chamber before finally getting into the one (1) million gallon clear well reservoir. The pumps for the various subsystems then lift the water into the transmission mains or lines. The suspended particles that settle in the clariflocculators are pumped onto a Slow Sand Filter bed for treatment.

3.3 Sampling

In this research work, stratified sampling method was used. Three sampling sites at the headworks were used to collect samples of raw water, settled water, filtered water and final water for analysis. However, results of analysis conducted by the headworks were also used. Before samples were taken, glassware and sampling bottles were washed with detergent and rinsed thoroughly with tap water. Subsequently, sampling bottles and glassware were rinsed using methylated spirit and rinsed with distilled water, after which they were sterilized finally with their lids fitted in an autoclave for twenty-five minutes at 120°C and allowed to cool before use. Moreover, glassware and sampling bottles were coded and neatly labelled. Analyses of samples were carried out at Kwanyako Water Treatment Plant Laboratory.

3.4 Equipment and Glassware

HACH MP 125 pH meter, 250ml conical flasks, Burette, Pipette, Wash bottle, HACH CO 150 conductivity/TDS Meter, Lovibond, Sampling bottles, Beakers, Autoclave, Clamp, HACH DR/2000 Spectrometer, Incubator, Analytical balance

3.5 Methods of Analysis

3.5.1 pH

The Hatch MP 125 pH meter which was calibrated before use, was rinsed with distilled water and the probe dipped into the beaker containing the raw water sample. The read button on the pH meter was pressed and allowed reading to stabilize before recording the pH value from pH meter. The procedure was repeated for settled, filtered and final water samples and their corresponding pH values recorded respectively.

3.5.2 Colour

The colour of the water is measured by means of colorimeter method of analysis. When testing for colour, 10ml of raw water was measured and topped to the 50ml mark in the test tube (lovibond cell). The lovibond cell containing the raw water was then placed into the right side compartment of the comparator in which a blank solution (distilled water) already placed in the left side compartment of the comparator served as control. The lovibond disc was used to determine the colour of the raw water by adjusting the disc to match with the colour of the blank.

34

The value of the colour of raw water was read on the disc and recorded. It is measured in Hazen Unit (H.U)

3.5.3 Turbidity

Using the absorptometric method, the standard programme number was entered into DR/2010 spectrometer and the wavelength set to display 'NTU TURBIDITY' UNITS. Two cells were obtained. One was filled with 25ml of deionised water (blank) and the other filled with 25ml each of raw, settled, filtered and final water samples. The blank was placed into the cell holder to tare (zero) the spectrophotometer. Each sample was placed into the cell holder one after the other and the results in Nephelometric Turbidity Units displayed. The value of each sample recorded.

3.5.4 Residual Chlorine

The quantity of chlorine in the final water was determined using a comparator.10ml of the final water was fetched into a test tube and one tablet of DPD placed into the test tube containing the sample and allowed to dissolve. After dissolving, the colour formed was then compared to the standard (blank) by moving the chlorine disc to match the colour of the blank and the corresponding value is recorded.

3.5.5 Total Alkalinity

The total alkalinity as expressed in terms of calcium carbonate was determined by titration using a standard solution of 2M sulphuric acid (H_2SO_4) with methyl orange as indicator. 50ml of water sample was measured by means of pipette and poured into a 250ml Erlenmeyer flask, two drops of methyl orange indicator added and mixed thoroughly by swirling. The burette content was titrated against with concentration of alkalinity was determined as

3.5.6 Total Hardness

50ml of each sample was measured into 250ml Erlenmeyer flask and two to three drops of ammonia (NH_3) buffer solution with pH of 10 was added to each sample. Few portions of Eriochrome Black-T are also added and the solution turns violet. 0.01M EDTA was pipetted and titrated against each sample with continuous stirring drop by drop until the violet colouration changed to blue black indicating that endpoint has been reached. The value was recorded and appropriate calculation made to obtain the actual value of total hardness. The procedure was repeated for each of the collected water samples.

Calculation: Total hardness (mg/l) as $CaCO_3$ = Titre value (EDTA) × 1000 ÷ Volume of sample used (ml).

3.5.7 Chloride

The chloride concentration of the sample was determined by titration. 10ml of each sample was measured into a250ml Erlenmeyer flask and two drops of potassium chromate indicator was added. 0.1N $AgNO_3$ was titrated against each sample with slow continuous swirling until the end point was reached. There was a colour change from yellow to pink. The procedure was repeated for each sample and the average titre value was calculated. Distilled water was also run through the procedure. Calculations were made to obtain the actual value for chloride concentration as mg/L chloride.

Chloride = (a-b)ml x N x 35.45 x 1000mg/L ÷ Volume (ml) of sample used

Where a = Titre value for sample,   b = Titre value for blank,   N = Normality of titrant (silver Nitrate)

3.5.8 Manganese hardness

Total hardness in water is due to the presence of Ca^{2+} and Mg^{2+} ions in the water. Magnesium hardness is obtained from the difference between the total hardness and calcium hardness. Magnesium hardness = total hardness – calcium hardness

3.5.9 Calcium hardness determination

50ml of the sample was pipette into a conical flask and 2ml of NaOH was then added and a few grains of murexide indicator were then added. The solution was then titrated with 0.01 EDTA solutions while shaking the conical flask to ensure constant mixing until a colour change from salmon to purple. The procedure was repeated and average titre value taken. Calcium hardness = volume of EDTA (ml) x 1000mg/L $CaCO_3^{2-}$ ÷ Volume of sample (ml)

3.5.10 Manganese determination

The HACH laboratory spectrophotometer (model DR/2500) was powered and the "HACH PROGRAM" selected from the main menu to display the specific programs. The specific program "295 manganese HR" was selected and the 'start' icon selected. A round cell was filled with 10ml of the sample, the content of one Buffer Powder Pillow Citrate type for manganese added, the cell capped and inverted gently to mix. The content of one Sodium Periodate Powder Pillow was again added to the sample cell, capped and inverted gently to mix [A violet colour developed in the presence of manganese] and the reaction timed for two minutes. Another round sample cell was filled with 10ml of the sample (this was the blank), placed into the cell holder and the 'zero' icon touched to display 0.0mg/L Mn. The sample was then placed into the cell holder, the stable result in mg/L Mn read and recorded.

3.5.11 Iron determination

Analysis of iron was done by using HACH (DR) 2000 spectrophotometer. The stored programme number for iron, 265 was entered. The wavelength dial was rotated till it displayed 510nm. Samples were poured into 25ml caveat, one pillow of FerroVer Iron Reagent Powder was added, swirled to mix and three minutes reaction time was allowed to elapse. The spectrophotometer was then zeroed with a 10ml of blank sample without a pillow addition. The prepared sample was inserted into the cell holder and the shield was closed. The stable result in mg/L Fe was read and recorded.

3.5.12 Bacteriological Analysis

Water contains a large number of microorganisms but the most important consideration is kind of microorganism present. The major characteristics of water are the absence of pathogens. Microorganism of interest to water bacteriology is the coliform group of bacteria. Coliform bacteria comprise many facultative anaerobic bacteria that develop red colonies with metallic sheen within twenty-four hours at 35° C on endotype medium containing lactose. MacConkey broth was used in testing for the presence of coliform bacteria based on colour change in the media using the multiple tube fermentation technique. A double strength MacConkey broth was prepared 40g to one litre of distilled water, stirred to dissolve completely. 10ml of the medium was dispensed into culture tube with inverted Durham tubes. The medium was inverted to ensure that the inverted were filled with the medium. The culture tubes were then sterilized in an autoclave at 121°C for fifteen minutes. The assumption for a large number of bottles takes the same format as the agar. After sterilization, the broth cooled and inoculated with 10 ml of the sample; it is then covered and incubated for 24 hours at37°C. The mouth of the sampling bottles, tap, media bottles containing broth and all other glassware were flamed before and after pouring samples into broth. This is done to kill any microorganism which is likely to cause contamination. The sampling bottles were inspected after 24 hours for gas production and tubes incubated for another 24 hours. After 48 hours of incubation, positive and negative were recorded. The positive ones were inoculated in brilliant green media as confirmatory test. They were incubated at37°C for 24 hours. After 48 hours incubation, negative results were recorded. If gas production was observed, the results were interpreted as follows:

❖ No gas produced, no colour change – Negative presumptive test

❖ Yellow colouration but no gas produced – Doubtful presumption test

❖ Yellow colouration with gas production – positive test.

3.5.13 Media preparation

3.5.13.1 Preparation of MacConkey Broth

The double strength broth was prepared and used to ensure that the smallest number of microorganisms (coliform) was detected by dissolving twice the mass (80g) of the MacConkey broth in 1 litre of distilled water, stirred to dissolve completely.10ml of the dissolved broth was pipette into each bottles fitted with inverted fermentation (Durham) tubes, capped and sterilized by autoclaving at 100°C for 15 minutes.

3.5.13.2 Preparation of Brilliant Green Bile Broth

40g of brilliant green bile powder was dissolved in 1 litre of distilled water. It was then brought to boil, distributed into bottles fitted with Durham tubes and sterilized by autoclaving at 121°C for 15 minutes.

CHAPTER FOUR

4   RESULTS AND DISCUSSIONS

4.1   RESULTS

TABLE 1: GDWS AND W.H.O GUIDELINE VALUES   ERROR! BOOKMARK NOT DEFINED.

Parameter	GDWS guideline values	WHO guideline values
pH	6.5 – 8.5	6.5 – 8.5
Colour (Hazen Units)	≤ 5	15
Turbidity (NTU)	≤ 5	5
Total Hardness (mg/L)	500	500
Manganese (mg/L)	0.1	0.1
Total dissolved solids- (mg/L)	1000	1000
Chloride (mg/L)	250	250
Iron (mg/L)	0.3	0.3
Residual free Chlorine (mg/L)	0.2	0.2
Sulphate (mg/L)	250	250

Table 1: Monthly mean values for 2009

PARAMETER		MONTHLY MEAN VALUES FOR 2009											
		JAN	FEB	MAR	APR	MAY	JUN	JLY	AUG	SEP	OCT	NOV	C
pH	RAW	7.125	7.150	6.975	7.025	7.100	7.100	7.260	7.300	7.150	7.100	7.100	7
	FINAL	6.975	7.300	6.550	6.475	6.575	6.725	6.94	7.050	7.075	6.800	7.775	6
COLOUR	RAW	110	106.2	120	101.2	195	190.5	160	66.25	43.75	195	198.7	7
	FINAL	4	4	4	4	4	6	4.2	4	4	4.5	4.75	4
TURBIDITY	RAW	22.2	17.07	17.03	19.62	43.15	35.01	25.66	16.27	9.967	50.15	54.43	1
	FINAL	0.5	3.713	o.32	0.407	0.6	0.85	1.13	0.825	0.585	1.15	0.7	0
RESIDUAL CHLORINE	FINAL	1.313	1.375	0.75	1	0.575	2.2	0.38	0.363	0.7	1.05	0.425	0
ALKALINITY	RAW	93.5	82	70	71.25	75.75	82.25	74.4	84.5	83.75	56.25	63	7
	FINAL	70	67	41	38.25	36.25	45.25	53.4	64.5	65	31.52	28.13	5
TOTAL HARDNESS	RAW	71	75.5	69.5	69	61	64	57.6	62	59	48.5	57	9
	FINAL	83	89.51	79.5	52.45	65	71.5	66.4	65.5	59.5	56.5	62.5	8
CALCIUM HARDNESS	FINAL	36.5	152	36.5	35.5	35.5	33	32.4	33.5	33.25	28	31.5	3
	RAW	44	204	46	36.5	38.5	37.5	35.6	38	37.75	34.5	36	4
MAGNESSIUM HARDNESS	FINAL	34.5	33.25	33	33.5	25.5	124	27.2	28.5	27.25	20.5	25.5	3
	RAW	39	38.5	33.5	33.5	29	34	30.73	27.5	27.75	22	26.5	3
CHLORIDE	FINAL	26.87	40.25	39	43	17	33.5	36	34.75	32	37.25	36.5	4
	RAW	29	43.5	41.13	45.25	18.88	35.13	38.8	36.75	34.75	40	39	4
IRON	RAW	0	0	0	0.525	0	0	0	0	0	0	0	0
	FINAL	0	0	0	0.075	0	0	0	0	0	0	0	0
TOTAL COLIFORM	RAW	0	0	0	0	0	0	0	0	0	0	0	0
	FINAL	0	0	0	0	0	0	0	0	0	0	0	0

Table 2: Monthly mean values for 2010

PARAMETER		MONTHLY MEAN VALUES FOR 2010											
		JAN	FEB	MAR	APR	MAY	JUN	JLY	AUG	SEP	OCT	NOV	D
pH	RAW	7.025	6.866	6.825	6.9	7.133	7.275	7.175	7.233	7.2	7.475	7.75	7
	FINAL	7.575	6.9	7.2	7.033	7.566	6.575	6.7	6.866	6.45	6.9	7.25	7
COLOUR	RAW	73.75	81.66	101.2	105	165	182.5	148.7	116.6	185	210.2	227.5	1
	FINAL	4	4.5	5.75	4	3.566	4.25	3.75	4.333	4	6.25	5	8
TURBIDITY	RAW	17.77	12.13	21.68	17.56	31.2	49.35	50.5	31.8	42.1	50.0	61	3
	FINAL	2.05	1.04	1.285	1.826	4.366	1.52	1.455	1.356	0.6	1.705	0.39	1
RESIDUAL CHLORINE	FINAL	1.162	1.21	0.467	1.033	0.47	0.687	0.687	0.456	0.685	1.75	1	1
ALKALINITY	RAW	74.74	77.33	50.25	78.66	71	69.5	71	74	63	58	71.5	8
	FINAL	69.25	59.33	35.75	46.66	44.66	29	39.5	45.33	38	33.25	40.5	6
TOTAL HARDNESS	FINAL	71	65.33	67	82.66	71.33	56.5	57	66	52	51.5	57	6
	RAW	82.5	83.33	76.75	106.6	103.3	60	62	68.66	55	60.5	60	7
CALCIUM HARDNESS	FINAL	36.5	34	35.5	58	32	29	30	34.66	27	25.25	30	3
	RAW	53.5	45.33	45	72.66	55.33	32	33.25	38	30	30	34	4
MAGNESSIUM HARDNESS	RAW	34.5	31.33	31.5	40	39.33	27	27	31.33	25	26.25	27	3
	FINAL	29	38	31.75	55.33	48	27	28.75	30.66	25	30.5	26	2
CHLORIDE	RAW	33.5	26.93	39	68.33	56	32.5	28.5	40.33	34.5	79.5	30.5	3
	FINAL	32	23.22	36.5	65.66	50	29	26	37.33	32.5	64.24	28.5	5
IRON	RAW	0	3.13	0	0	0	0	0	0	0	0	0	0
	FINAL	0	0.13	0	0	0	0	0	0	0	0	0	0
TOTAL COLIFORM	RAW	0	0	0	0	0	0	0	0	0	0	0	0
	FINAL	0	0	0	00	0	0	0	0	0	0	0	0

Table 3: Monthly Mean values for 2011

PARAMETER		MONTHLY MEAN VALUES FOR 2011											
		JAN	FEB	MAR	APR	MAY	JUN	JLY	AUG	SEP	OCT	NOV	DEC
Ph	RAW	7.5	7.2	7.316	7.1	7.333	7.575	7.62	7.5	7.5	7.35	7.3	7.25
	FINAL	7.55	7.4	7.3	6.95	7.025	7.025	7.2	7.725	6.725	7.025	7.6	6.875
COLOUR	RAW	76.25	69.75	83.33	157.5	246.2	151.2	223.7	110	211.2	20	161.2	86.25
	FINAL	4	3.25	3.833	4	4	4.5	4.25	3.75	4	4.5	4	3.5
TURBIDITY	RAW	19.675	46.1	16.53	40.35	55.25	45.47	62.47	24.2	63.92	41.2	51.57	18.37
	FINAL	0.8275	0.7	0.54	0.44	0.607	1.357	0.867	0.825	0.543	0.685	0.632	0.755
RESIDUAL CHLORINE	FINAL	0.9562	0.625	0.766	0.825	1.2	0.637	1.425	1.15	1.3	1.325	0.725	0.7
ALKALINITY	RAW	90.25	57.5	78	70	60.75	67.5	61.25	71	63.5	54	65.5	75
	FINAL	74.5	42.5	56.66	46.5	37.75	37	34.75	56	32.5	31.75	41.5	54.5
TOTAL HARDNESS	FINAL	61	40	42	50	54.5	57	55.75	61.5	55.75	50	55.5	70.5
	RAW	61	50	47.33	55	64.25	59	58	77	60	58	64	77
CALCIUM HARDNESS	FINAL	34.5	34	35.33	32	32	29.5	28.75	34	30	29	32.25	38
	RAW	42.5	44	41.33	39	39	30.5	35.25	47	35	33	38.5	41.5
MAGNESSIUM HARDNESS	FINAL	26.5	6	4	18	22.5	27.5	27	27.5	25.75	21.5	23.25	32.5
	RAW	28	6	6.666	16	25.25	28.5	17.75	30	25	25	25.5	35.5
CHLORIDE	RAW	33.75	41	37.5	42.5	41	34.5	35.25	109	30.5	29.75	33	36.65
	FINAL	31.5	38	35.83	40.5	39.33	31.75	31.25	24.73	27.25	25.75	29.75	34.5
IRON	RAW	0	0	0	0	0	0	0	0	0	0	0	0
	FINAL	0	0	0	0	0	0	0	0	0			0
TOTAL COLIFORM	RAW	0	0	0	0	0	0	0	0	0	0	0	0
	FINAL	0	0	0	0	0	0	0	0	0	0	0	0

4.2   DISCUSSIONS

4.2.1   pH

From the graph below, it can be seen that there are no significant changes in monthly mean values of pH. Most of the parameters were within stipulated GDWS and WHO guideline values of 6.5 – 8.5. With the exception of April 2009 and September 2010 which fell below the guideline value, indicating under dosing of lime for pH correction, most of the final water at the headworks was non corrosive. This is because corrosive water (water with low pH) reacts with the metallic part of the pipeline causing quality deterioration of final water, stains plumbing fixtures as well as its associated health impacts.

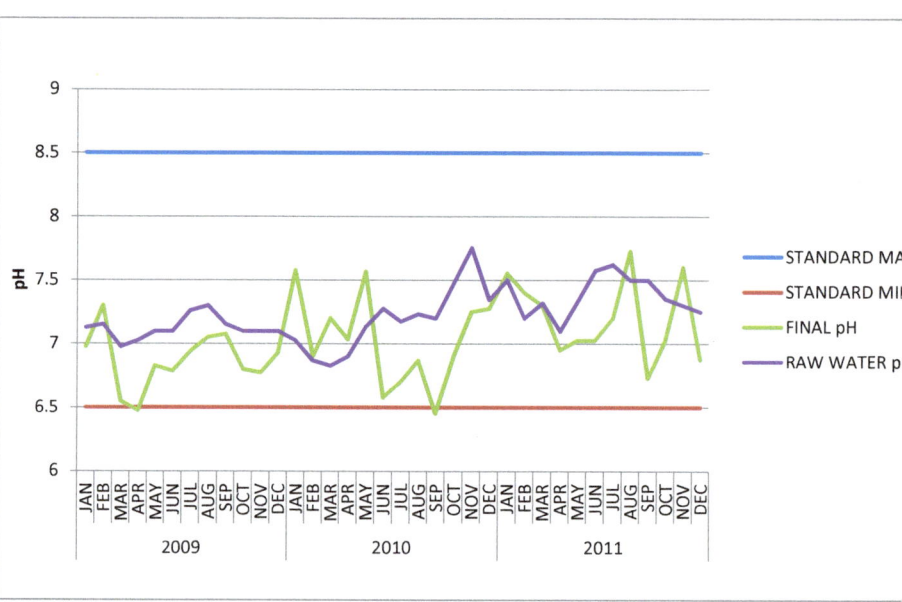

Figure 8: A graph depicting pH variation against GDWS and WHO guidelines

4.2.2 COLOUR

From the graph below, it can be observed that all the parameters were within the WHO guideline value of 15 H.U. But regarding GDWS guideline value of ≤ 5 H.U, most parameters were within standard except June 2009, October and December 2010 monthly mean values which exceeded GDWS guideline value. This may be due to decaying organic matter of vegetation, high iron and manganese build-up at point of raw water abstraction, algae growth, economic activities along raw water source and seasonal changes in raw water characteristics as well as dissolved or colloidal particles. In respect of colour, raw water with relatively high values of 202.5 in 2009, 286.2 in 2010 and 246.2 Hazen Units in 2011 were considerably reduced to 4, 5 and 4 Hazen Units respectively of final water colour which showed tremendous plant performance of raw water treatment.

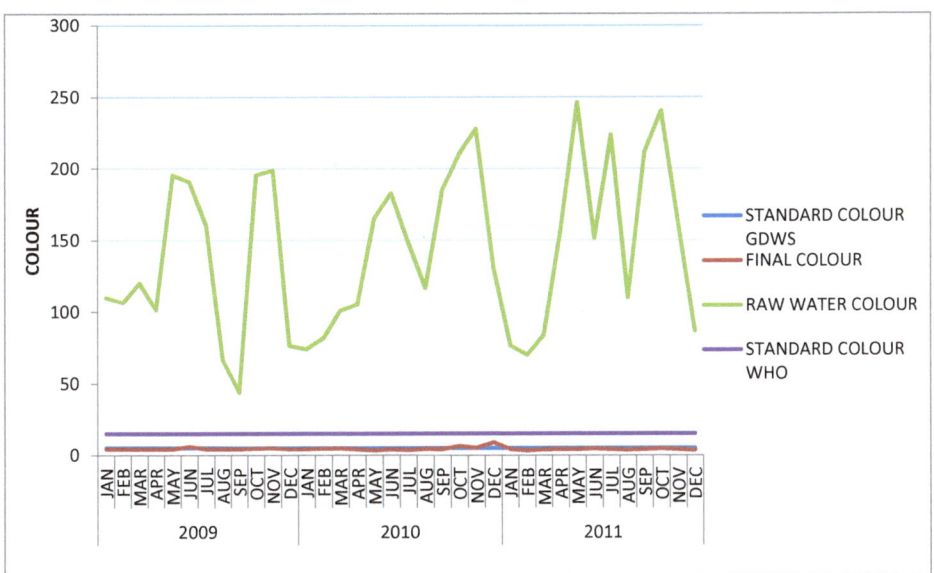

Figure 9: A graph of apparent colour against GDWS and WHO guidelines

4.2.3 TURBIDITY

With reference to fig. 3, it can be observed that final turbidity values over the period were within GDWS and WHO guideline values. The highest turbidity mean values of raw water recorded in 2009, 2010 and 201 are 54.43 NTU, 84.9 NTU and 74.47NTU respectively with their corresponding final water mean values of 0.7 NTU, 1.705 NTU and 0.867 NTU. This indicates plant remarkable performance of removing dissolved, suspended and particulate matter over the period.

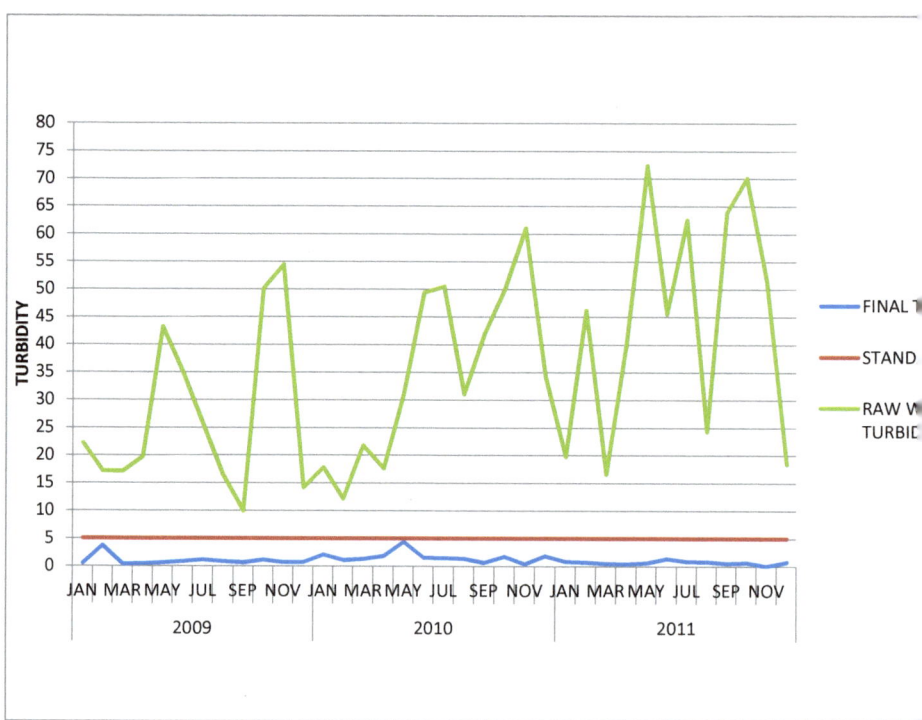

Figure 10: A graph Turbidity in comparison with GDWS and Who guidelines

4.2.4 RESIDUAL CHLORINE

From fig. 4, residual chlorine levels in final water at the headworks over the period were relatively high above the acceptable limit with exception of December, 2009 which fell below the standard. High dose of residual chlorine at the headworks was recorded especially, in June, 2009 due to changes in raw water characteristics.

Similarly, microorganisms, pathogens and other contaminants which can spread disease to humans during distribution are drastically rendered inactive and subsequently damaged in order to make the water biologically sterile and safe for consumption. This role reduces high levels of residual chlorine to acceptable standard before the water reaches the consumers.

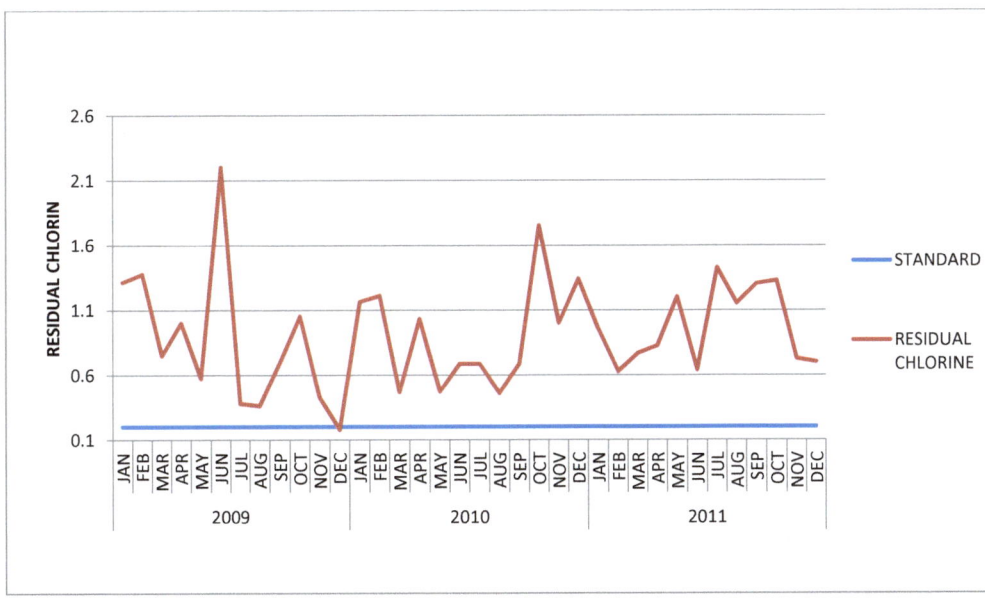

Figure 11: A graph of Residual Chlorine against GDWS and WHO guidelines

4.2.5 TOTAL HARDNESS

Water Hardness can be classified as soft when it ranges from (0–50 mg/L CaCO$_3$), moderately soft (50 – 100 mg/L CaCO$_3$), slightly hard (100 – 150 mg/L CaCO$_3$), moderately hard (150 – 200 mg/L CaCO$_3$), hard (200 – 300 mg/L CaCO$_3$) and very hard (over 300 mg/L CaCO$_3$). With reference to fig. 5, it can be observed that between 2009 and 2011, the hardness of final water treated at the headworks can be described as soft and moderately soft with minimum mean value of 47.33 mg/L and maximum mean value of 87.51as Calcium and magnesium. In 2010, hardness of finished water can be classified as slightly hard with recorded highest mean value of 106.6 as Calcium and magnesium. Plant performance regarding hardness of final water had been efficient since calcium and magnesium contents in raw water have been treated or reduced from 288 mg/L to52.45 mg/L which an acceptable standard for domestic purposes.

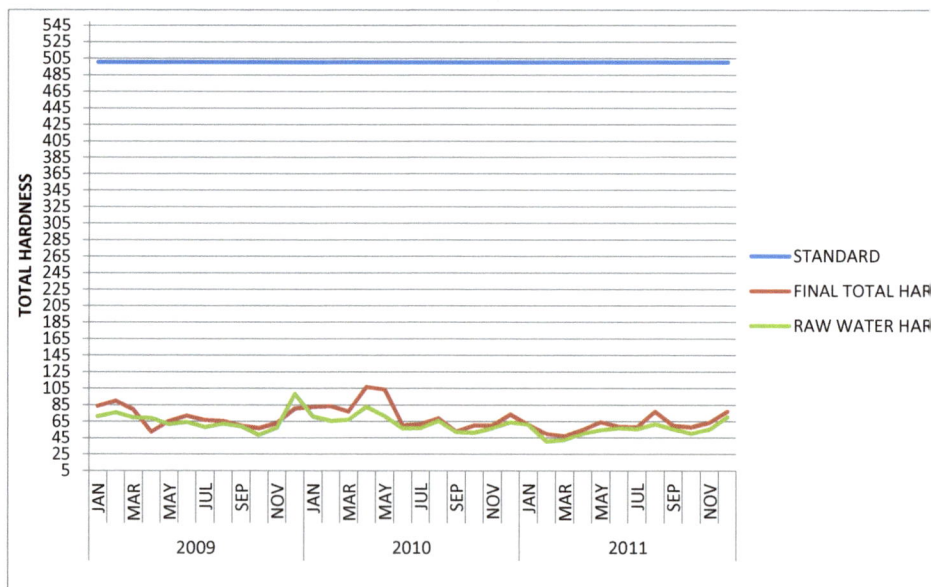

Figure 12: A graph of Total Hardness against GDWS and WHO guidelines

47

4.2.6 CHLORIDE

All the monthly mean values of the parameter were far below GDWS and WHO guideline values of 250 mg/L. The monthly mean values recorded were 116 mg/L, 79.5 mg/L and 109 mg/L for 2009, 2010 and 2011 of raw water respectively with their corresponding final values of 34.74 mg/L, 26.93 mg/L and 29.75 mg/L. Chloride in drinking water may originate from sources such as sewage, industrial effluents, urban runoffs and sea water intrusion due to over pumping. The low values of chlorides indicate plant efficiency.

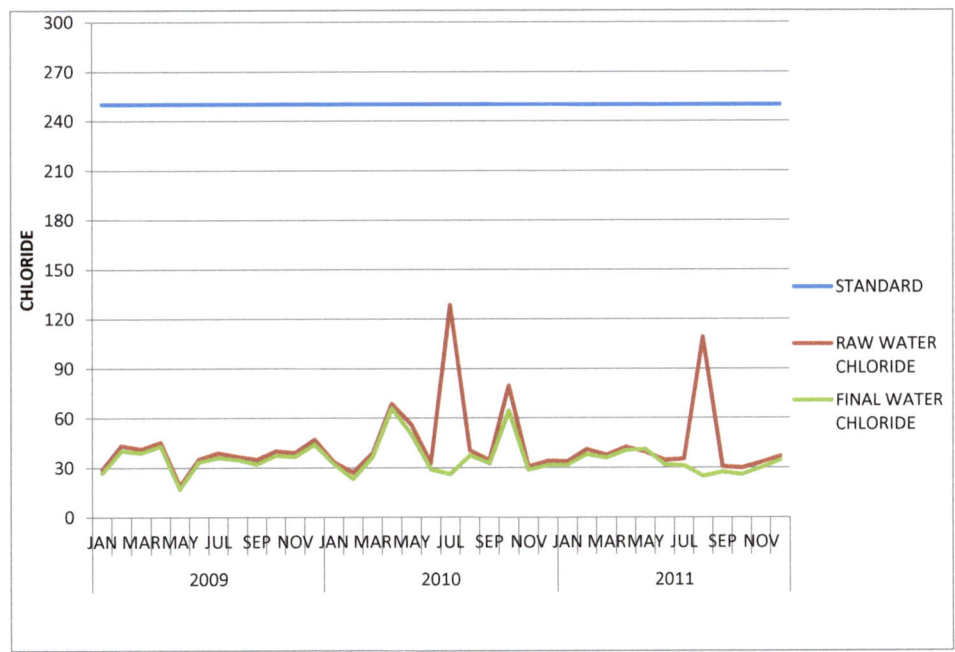

Figure 13: A graph showing Chloride comparison with GDWS and WHO guidelines

4.2.7 TOTAL IRON

The monthly mean values of iron over the period were relatively lower than GDWS
and WHO guideline values of 0.3mg/L. Final water recorded an average value of
0.075mg/L in April 2009, but an increase of 0.055mg/L was recorded giving rise to
0.13mg/L in February 2010 while in 2011, there were no incidence of iron.

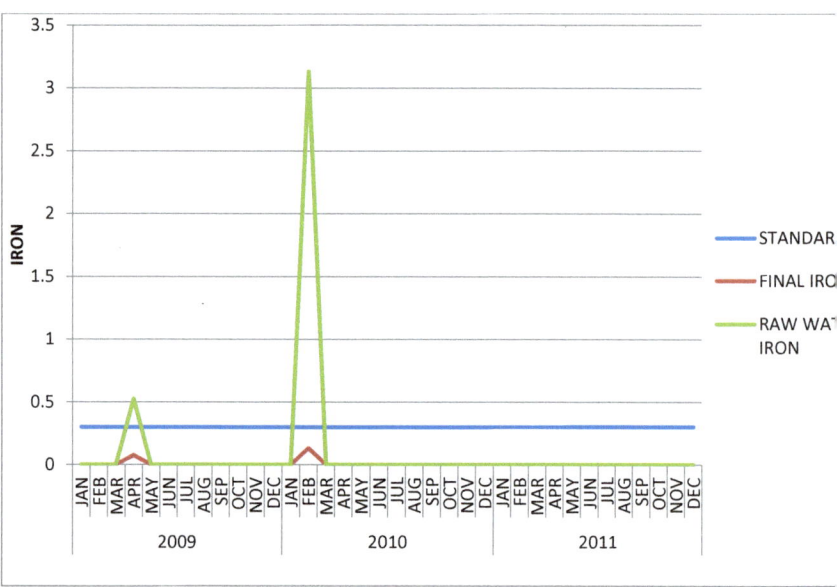

Figure 14: A graph representing Total Iron comparison with GDWS and WHO guidelines

4.2.8 TOTAL COLIFORM

It can be observed from the results that total coliform recorded over the period was
zero which indicated that there were no biological growths in final water. Coliform
determination of raw and final water gives indication of pollution source of water.
Presence of these indicative organisms is evidence that the water has been polluted
and their absence also shows unpolluted water source.

CHAPTER FIVE

5   CONCLUSION AND RECOMMENDATION

5.1   CONCLUSION

By the results obtained from the monthly analysis, it was realised that most of the parameters analysed fell within the GDWS and the WHO guideline values stipulated with the exception of residual chlorine which had been duly discussed. Regardless of high levels of residual chlorine recorded at the headworks, it can be generally concluded that the plant performed creditably well against all the parameters with reference to the standards. Therefore, plant efficiency was relatively high and results also reliable in producing good quality water for consumption.

5.2   RECOMMENDATION

Based on the result of analysis, it is being recommended that;

❖ The chlorinator should be calibrated periodically to give accurate results

❖ Proper plant operation and maintenance programme should be encouraged at the headworks to permit continuous functioning of plant.

❖ Mechanical equipment should be periodically disassembled, systematically inspected, lubricated and overhauled.

❖ The water treatment plant facilities should be managed and controlled by competent personnel.

❖ Plant records should be available, proper storage of chemicals must be maintained and plant cleanliness encouraged.

REFERENCES

A., p.-U., Bos, R., & F.andbartram, J. (2008). Safer Water,Better Health,Cost,Benefits and Sustainability of Interventions to Promote Health. Geneva.

Anipa, H. (2001). Performance Evaluation of Kpong Water Treatment Plant. KNUST,Ghana: MSc.Theisis for the Msc Programme in WESE.

Anor G.A, S. M. (2012, Feb. 6). Kwanyako Treatment Plant. (O. J. Odame, Interviewer)

ASCE, & CSSE, A. a. ((1971).). Water Treatment Plant Design. pp. 1 – 2.

AWWA, J. (1968). Quality Goals for Potable Water. Quality Goals for Potable Water , 60(12):1317.

Boateng, P. D. (2012). Process and Quality Control in Water Supply, lecture notes.

Boateng, P. D. (2010). Process and Unit Operation, Lecture Notes. Ucc, Ghana.

Bryant, E., & Fulton, G. B. (1992). Disinfection Alternatives for Safe Drinking Water, Hazen and Sawyer Environmental Engineers and Scientists. New York: Van Nostrand Reinhold Publishers.

Buamah, R. ((2010)). "Principles of Filtration". Lecture notes. KNUST, Ghana.

Duncan, A. E. (2011). Water Treatment Processes, Lecture Notes. Ucc, Ghana.

Flentje, M. (1965). "Standard Methods for the Examination of Water and Wastewater". New York: American Public Health Association, Inc.

Flentje, M. E. (1985). Treatment Plant Control. In P. D. Boateng, Process and Quality Control in Water Supply, Lecture notes (pp. 464-472).

Franson, M. A. (1984). Standard Methods for the Examination of Water and Wastewater. USA: American Public Health Association, 16th Edition, American Water Works Association, WPCF.

GWCL, Ministry of Works and Housing. (2008). Annual Diary. Accra.

Howard S. P., D. R. (1985). Environmental Engineering. New York: McGraw-Hill book Company.

Huisman, I. ((1986)). "Rapid and Slow Sand filtration", Lecture notes. IHE, DELFT, EE 178/86/1, 2002.

J.A., N. (2000). Basic Environmental Technology:water supply,waste management and pollution control 3rd Edition. New Jersey: Prentice Hall,Cranford.

James Montgomery Consulting Engineers Inc. (1985). Water Treatment Principles and Design, J. w.–2.

James, M. M. (1985). Water Treatment Principes and Design. Canada, New York, USA: John Wiley and Sons Inc. .

L.R., M. (1962). Water Supply and Treatment, 9th Edition. Washington D. C.: Bulletin211, National Lime Association.

Masel, I. R. ((1996).). Principles of adsorption and reaction on solid surfaces. pp. 247.

Montgomery, M., & Bartram, J. E. (2009). increasing functional sustainabililty of water and sanitation supplies in rural sub-saharan Africa. Journal of Environmental Engineering Science.126(5) , 1017-1023.

Peavy, H. S., & Rowe, D. R.-H.

Salvato, J. (1992). Environmental Engineering Sanitation,4th Edition. John Wiley and Sons Inc. pp. 331-345.

Sharma, S. K. ((1997)). Iron Removal in Water Treatment by Adsorption of Iron (II) onto filter media., MSc. Thesis presented to IHE Deft. Netherlands.

Spellman, F. (1999). Spellman's Standard Handbook for Wastewater Operators, Vol.1. Lancaster: Technomic Publications.

T.H.Y., T. (1998). Principles of Water Quality and Control,5th Edition. Oxford: Butterworth-Heinemann,pp.1.

W.H.O/UNICEF. (2010). Joint Monitoring Programme for Water Supply and Sanitation. France: 2010 update.

www.ingramcontent.com/pod-product-compliance
Lightning Source LLC
Chambersburg PA
CBHW040842180526
45159CB00001B/287